AI TECHNOLOGY
LIFESTYLE
PREDICTIONS FOR THE BODY AND MIND

PATRICK JACKSON

Ordering Information:

Prime Seven Media
518 Landmann St.
Tomah City, WI 54660

Printed in the United States of America

W e humans can Communicate through telepathy. Telepathy is the mind, Telecommunications, It is a way of sending high frequency signals thinking, Between you and another person. In the future we will be using AI technology, To pick up signals From telepathy Without having to use a mobile phone, A I technology will make this possible, In the future we can have a minicomputer built or fitted onto our bodies. Or something we can wear. Like a small watch So A I computer Will be able to. Pick up the signals from the body And transfer it To a more powerful computer So that we can, Communicate to each other, And the system can also Diagnose Problems With our health body or mind, a Minicomputer Is a little bit like, Having a doctor Walking around with you, This small computer Will have a small. Microprocessor built in to keep a check on your health, And process information Problems about your health. You could have a dream. Whether your dream is a good dream or bad dream or even a nightmare. And this could be picked up by AI computer. A specialist could actually analyse that dream could be recorded on video, From your AI mobile Phone System, and Send it to a computer, so other people can look into it, And certify and examined, a person mind, Maybe that dream is putting stress on you, making you sick. Telepathy is a high frequency. Mind. Telecommunications. That is a communication signal. Transferred Between one person to another person, In the future AI computer will be able too Pick up signals Of telepathy, And convert them Into signals That can be sent by a human To a another person Through telepathy, We're in 2024. And there's not a computer That can convert Telepathy signals From a human being To another human being!, Many years ago. The inventors.

Developed high frequency signals. For radio, television, radar, we use those signals, for our advantage today. In the time to come, Telepathy communications Will be standard for Everyone to use telepathy. A I communications Technology. Have you ever thought? Why the body Use electricity? The body stores a small amount of electricity, For the brain to process information and send, High frequency Signals To other parts of the brain, to control The body and mind. What is the difference between the brain and the mind? the brain controls a person's movements, emotions and various bodily functions, the mind alludes to a person's morality, reasoning and understanding. The brain is a physical organ. It can be touched or seen. However, the mind is intangible. It can either be touched nor seen. What is the difference between the body and mind? According to Better Up, your mind is your thinking mind that is responsible for your beliefs, thoughts, an actions body. Your body is the physical aspect of yourself that carries you through life and allows you to experience the world through your 5 senses. Why the body use electricity? According to University of Maryland, Baltimore graduate programmes, electricity is required for the nervous system to send signals throughout the body into the brain, making it possible for us to move, think and feel. Can the brain control electricity systems in our body? According to Re an electrical signal can be generated from your brain where they travel to different areas of your body, or it can work the other way around. In the time to come. A small microprocessor chip. can be fitted to the body or the head, or it can be, a wearable device like a watch you can you can wear on your hand, Or it could be fitted too the head, So that electrical signals can be picked up easy. For the computer to do a diagnostic test On the body And the brain? To see if that person, Body is healthy Or sick?. In the time to come, Wearable devices, of high technology watch, that Will become a standard, For A I. Communication technology. To do a diagnostic test On your body, And if that person is getting a heart attack? Or a stroke, The computer Will send information To your doctor's. To let them know Your life is in danger And also The AI computer. Could automatically call the ambulance, To your location, Where you are, So that the doctors. Can examine you, And save your life. We humans suffer from a lot of Sickness, Illness, And diseases, That I will be highlighting, In this storyline! Why does the brain suffer from loss of memory? According to Cleveland Clinic, acute memory loss, commonly known as amnesia, This usually happens because of a sudden illness, injury, or other events that disrupt your memory process is progressive memory loss. This is memory loss that happens gradually. It's sometimes a symptom of a degenerative brain disease. What is the next word for Alzheimer's? According to Alzheimer's Society of Canada, while the terms Alzheimer's disease and dementia are often used interchangeably, it's important to know the difference between the two. Dementia is not one specific disease. Rather, it's an umbrella term for a set of symptoms caused by physical disorders affecting the brain. What is Alzheimer's? According to Centres for Disease Control and Prevention, Alzheimer's disease is the most common type of dementia. It is a progressive disease, beginning with mild memory loss and possibly leading to loss of the ability to carry on a conversation and respond to the environment. Alzheimer's disease involves parts of the brain that control thought, memory and language. What is dementia? According to Centres for Disease Control and Prevention, dementia is not a specific disease but is rather a general term for the impaired ability to remember, think or make decisions that interferes with doing everyday activities. Alzheimer's disease is the main type of dementia, that affects mostly older people, It is not a part of normal ageing process, Is there a cure for Alzheimer's? According to Alzheimer's Society, there is no cure for Alzheimer's disease yet. However, with the right care and support, it is possible for someone with Alzheimer's to live as long as possible. Describe Parkinson's disease. According to Mayo Clinic overview, Parkinson's disease is a progressive disorder that affects the nervous system and the parts of the body controlled by the nerves. Symptoms start slowly, The first symptom may be a barely noticeable tremor in just one hand. Tremors are common, but the disorder may also cause Stifel slowing movement. Can Parkinson's disease be cured? According to Mayo Clinic, Parkinson's disease can't be cured, but medicines can help control the symptoms, often dramatically. In some more advanced cases, surgery may be advised. Your healthcare team also may recommend lifestyle changes, especially ongoing aerobic exercise. What is senile dementia? According to sciencedirect.com, the term senile dementia was used for many years to describe older individuals who suffered from cognitive decline, particularly memory loss. This term actually reflects a long history of not understanding dementia, its causes, or its treatment. How can split personality disorder, be treated? Here's a summary from mind talking therapy, one talking therapy, to accessing therapy 3. Choosing a therapist for dissociative disorders, and five medication. What is a split personality disorder?

According to HSBC, sociated with entity disorder DID used to be called multiple personality disorder. Someone diagnosed with DID may feel uncertain about their identity and who they are. They may feel the presence of other identity's, each with their own names, voices, personal histories and mannerisms. What does the word psychotic mean? This is the definition of psychotic relating to be affected with psychosis. Describe the word psychosis. According to National

Institute of Mental Health, National Institutes of Health. Psychosis refers to a collection of symptoms that affect the mind where there has been some loss of contact with reality. During an episode of psychosis, a person's thoughts and perceptions are disrupted and they may have difficulty recognising what is real and what is not. What is a Mental breakdown? According to Health Direct, a nervous breakdown, also known as a mental health crisis or mental breakdown, describes a period of intense mental distress people feel they're having. A nervous breakdown may be experiencing overwhelming anxiety, paranoia, or thoughts of self harm. Describe. Schizophrenia. According to National Institute of Mental Health, National Institutes of Health. What is schizophrenia? How many different brain diseases are there? According to Brain Foundation, there is no one answer to the question what is brain disease? Because many different conditions can affect the brain and different conditions vary greatly in symptoms, severity, diagnosis and treatment. There are over 600 diseases that can affect the nervous system. Schizophrenia is a serious mental illness that affects how a person thinks, feels and behaves. People with schizophrenia may seem like they have lost touch with reality, which can be distressing for them and for their family and friends. What is paranoid schizophrenia? According to Cleveland Clinic by Confused by Provision, because paranoia commonly happens with schizophrenia. Paranoia is a pattern of behaviour where a person feels distrustful and suspicious of other people and acts accordingly. Delusions and hallucinations are the two symptoms that can cause paranoia. What is Brain Damage? According to WebMD, brain damage is an injury that causes the destruction or deterioration of brain cells. How would a person know they've got brain damage? According to Mayo Clinic, symptoms headache, nausea or vomiting, fatigue or drowsiness, problems with speech and dizziness or loss of balance. Describe psychosis of the mind. According to nationalistic mental, national cyber collection of symptoms that affect the mind that he last activity code of psychosis, a person's thoughts and perceptions are disrupted and they may have difficulty recognising what's real and what is not. What is the human soul? According to Wikipedia many witches in philosophical this spiritual essence of a person which includes ones identity, personality and memories. An immaterial aspect or essence of a living being that is being able to buy physical. the human spirit is an extraordinary force within each individual, embodying A remarkable array of qualities and capabilities, From resilience to creativity, empathy to courage, the humour is its incredible potential in various aspects of life. According to LinkedIn, the human spirit, his the next before switch visual, embodying A remarkable array of qualities and capabilities from resilience to creativity, empathy to courage, The Human Spirit exhibition, various aspects of faith. What is diabetes? According to World Health Organisation, diabetes is a chronic disease that occurs either when the pancreas does not produce enough insulin cannot effectively reduces. Insulin is a hormone that regulates blood glucose. People of all ages can develop type one diabetes. If you have type one diabetes, your pancreas doesn't make insulin or makes very little insulin. Insulin helps blood sugar enter the cells in your body for use as energy. Without insulin, blood sugar can't get into cells and builds up in the bloodstream. Describe type 2 diabetes. According to NHS type 2, diabetes is a common condition that causes the level of sugar glucose in the blood to become too high. It can cause symptoms like excessive thirst, needing to pee a lot and tiredness. Many people have no symptoms. It increases your risk of getting serious problems with your eyes, feet, heart and nerves. What is blood pressure? According to Wikipedia, blood pressure is the pressure of circulating blood against vessels. What is breast cancer? According to the Centres for Disease Control and Prevention, breast cancer is a disease in which cells in the breast grow out of control. There are different kinds of breast cancer. The kind of breast cancer depends on which cells in the breast turn into cancer. Most breast cancers begin in the ducts or lobules. How can artificial intelligence help cure breast cancer and radiotherapy? According to Institute of Cancer Research demonstrating how I can improve the accuracy of diagnosis, they applied machine learning to molecular data and gene sequences from breast hammers. People trained AI model to fight types of disease with particular patterns of response to treatment. AI technology will be able to use. Diagnostic test to detect these illness, AI technology will be able to run a diagnostic test on our illness? According to LinkedIn, by analysing vast amounts of data from various sources such as medical images, lab results and patient history, AI algorithms can identify patterns and anomalies that may not be detected by human doctors. This enables early detection of diseases and reduces the chances of misdiagnosis. What is Leukaemia?

According to Cancer Council, what is leukaemia, leukaemia's or leukemias? US spelling accounts is of the white blood cells which begin in the bone marrow. Leukemias are grouped in two ways, the type of white blood cell affected, lymphoid or myeloid, and how quickly the disease develops and gets worse. Is there a cure for leukaemia? According to an extent all across the globe, it may not need to have treatment straight away. When you're diagnosed with chronic lymphocytic leukaemia, you usually be seen by a doctor or nurse regularly. What hopes is there for leukaemia patients?

According to BBC, personalised treatment for the most common form of adult leukaemia patients survive for longer and stay in remission, a study has found. A trial led by the University of Leeds found therapy could be individualised for each patient by using regular blood tests to monitor their response. What is cancer? According to NHS, cancer is a condition where cells in a specific part of the body grow and reproduce uncontrollably. The cancerous cells can invade and destroy surrounding healthy tissue, including organs. Cancer sometimes begins in one part of the body before spreading to other areas. How many different type of cancer is there? According to Cancer Research UK, but vary in some ways because body organs do very different things. For example, nerves and muscles do different things, so the cells have different structures.

There are more than two hundred types of cancer, and we can classify cancers according to where they start in the body, such as breast cancer or lung cancer. Could there be a cure for cancer in the future? According to Macmillan, Cancer Support for cancer that is treatable but not curable. Advances in treatment could transform cancer into a long term health condition. In addition, there is some justifiable excitement about vaccines. So yes, we will cure some cancers, but we won't cure the concept of cancer. What is skin cancer? According to Mayo Clinic overview Skin Cancer, the abnormal growth of skin cells most loops on supposed to the This common form of cancer can also occur on areas of your skin not ordinarily exposed to sunlight. There are three major types of skin cancer, basal cell carcinoma, famous cell carcinoma and Melanoma. Can a person get brain cancer? According to NHS Brothers can affect Page, including children, although they tend to be more common in older adults. More than 12,000 people are diagnosed with a primary brain tumour in the UK each year, of which about half a cancerous many others are diagnosed with a secondary brain tumour. Can a person get throat cancer? According to Yale Medicine, throat can break more men than women, likely due to the fact men use alcohol. While people over the age of 55 or at highest risk for developing throat cancer, it's growing more common in younger people due to the increased prevalence of HPV. Can a person get eye cancer? According to Cancer Research UK, Kissack ocular causes ocular is the medical name of the eye. Eye cancer is very rare. What is bowel cancer? According to Cancer Research

UK, bowel cancer is also called colorectal cancer. It affects the large bowel, which is made-up of the colon and rectum. Cancer is when abnormal cells start to divide and grow in an uncontrolled way. The cells can grow into surrounding tissues or organs and may spread to other areas of the body. What is prostate cancer? According to Mayo Clinic, prostate cancer is cancer that occurs in the prostate. The prostate is a small walnut shaped gland in males that produces the Seminole fluid that nourishes and transports sperm. Prostate cancer is one of the most common types of cancer. What is blood cancer? According to Blood Cancer UK, blood cancer is a type of cancer that affects your blood cells. Leukaemia, lymphoma and myeloma are some of the most common types of blood cancer. There are also types called MPN's, NMDS. Blood cancer is caused by changes, mutations in the DNA within blood cells. What type of disease is it that makes you lose your hair? According to National Institute of Arthritis and Musculoskeletal and Skin Diseases, Alopecia Aria, Theresa disease that happens when the immune system attacks hair follicles and causes hair loss, Hair follicles are the structures in skin that form hair. While hair can be lost from any part of the body, Alopecia arietta usually affects the head and face. How can we protect the body from diseases? Here's a summary from God site disease She eat healthy, get adequate sleep, exercise, manage stress, receive age-appropriate vaccinations, don't smoke and be sceptical. What are diseases? According to Wikipedia, the Taylor abnormal conversely effects that all part of an Organism and is not immediately due to any external injury. How can we control a disease in the body? Here's a summary from VI. Among the sensible actions you can take, keep immunizations up-to-date, wash your hands often, prepare and handle food carefully, use antibiotics only for infections caused by bacteria, and more. How can pneumonia be cured? According to Asthma Plus Lung UK, pneumonia can be serious, so it's important to get treatment quickly. The main treatment for bacterial pneumonia is antibiotics. You should also rest and drink plenty of water. If you're diagnosed with bacterial pneumonia, your doctor should give you antibiotics to take within four hours. What is a blood clot? According to Wikipedia, a thrombus record clot is final product of the blood coagulations step in haemostasis. What are varicose veins? According to NHS in full, varicose veins are swollen and enlarged veins – usually blue or dark purple - that usually occur on the legs. They may also be lumpy, bulging or twisted in appearance. The symptoms are heavy include: aching heavy and uncomfortable legs. How does people get asthma?

Asthma more often starts during childhood when your immune system is still developing. Multiple factors may work together to cause it, such as things in the environment called allergens that affected you as a baby or young child, which may include cigarette smoke or certain germs. Viral infections that affect breathing. According to NHS, this currently no cure for asthma, but treatment can help control the symptoms so you're able to live a normal, active life. Inhalers, which are devices that let you breathe in medicine, are the main treatment. Tablets and other treatments may also be needed if your asthma is severe. What is H I V? According to Wikipedia, the Human immunodeficiency viruses are two species of lentivirus that infect humans, How can H I V be cured? According to Centres for Disease Control Prevention, H I V from body healthy show it with a treatment, people can get the virus under control within six months. H I V treatment do not prevent transmission to other people. How does a person catch gonorrhoea? According to the Centres for Disease Control and Prevention, gonorrhoea is transmitted through sexual contact with the penis, vagina, mouth or anus of an infected partner. Ejaculation dose Not have to occur for gonorrhoea to be transmitted, from mother to baby. What is autism? According to Centres for Disease Control and Prevention, autism spectrum disorder, ASD is a developmental disability caused by differences in the brain. People with ASD often have problems with social communication and interaction, and restricted or repetitive behaviours or interests. People with ASD may also have different ways of learning. Moving or Paying attention? How can autism be cured? According to NHS, autism is where your brain develops differently to non-autistic people. It is not an illness and there is no cure. If you're autistic, a GP or local autism team may suggest approaches that can help you to develop daily living skills. What is bone cancer? According to Cleveland Clinic, bone cancer is the term for several different cancers that develop in the bones. When cancer cells grow in a bone, it can harm normal bone tissue.

The type of cell and tissue where cancer begins determines the type of bone cancer.

Cancer, that form in the bone itself are called primary bone cancers. How can bone cancer be treated? According to NHS, surgery to remove the section of cancerous bone, it's often possible to reconstruct or replace the bone that's been removed, But amputation is sometimes necessary.

Chemotherapy – treatment with powerful cancer – Killing medication Radiotherapy - where radiation is used to destroy cancerous cells. MND is still incurable, but it is not untreatable, as many symptoms can be managed. The drug riluzole – available on the pharmaceutical Benefits scheme- has been demonstrated in clinical trials to prolong survival by several months in the milder phase of the disease for longer life. Why do we have headaches? According to Mayo Clinic Chemicals, The Chemical activity in your brain, the nerves or blood vessels surrounding your skull or the muscles of your head and neck or(some combination of these factors) can play a role in primary headaches.

Some people may also carry genes that make them more likely to develop such headaches. why does people get migraine headaches? Some possible triggers include the following: stress and other emotions. Biological and environmental conditions, such as hormonal shifts or exposure to light or smells. Fatigue and changes in one's sleep patterns. I mentioned some illness we have. And how to understand them. And cure them. By seeing your doctor. Or medical professionals. What is a brain haemorrhage? According to New York Presbyterian, what is a brain bleed or brain haemorrhage? A brain bleed, also known as a brain haemorrhage, refers to bleeding between the brain tissue and the skull or inside the brain tissue. This is a life-threatening condition that requires immediate medical attention. How can a person get Sciatica? According to Mayo Clinic, sciatica occurs when the nerve roots to the sciatic nerve become pinched. The cause is usually a herniated disc in the spine or an overgrowth of bone, sometimes called bone spurs, on the spinal bones. More rarely, A tumour can put pressure on the nerve. Can sciatica be treated? According to Web MD, how do you get rid of sciatic nerve pain? You can usually treat a mild case of sciatica yourself with a combination of heat, ice over the counter, pain medication, and stretching and strengthening exercises. For more severe pain, your doctor might recommend physical therapy, steroid injections, or surgery. What is glaucoma? How can people get glaucoma? According to NHS, it's usually caused by fluid building up in the front part of the eye, which increases pressure inside the eye. Glaucoma can lead to loss of vision if it's not diagnosed and treated early. It can affect people of all ages, but is most common in adults in their 70s and 80s. How can glaucoma be treated? According to Mayo Clinic, glaucoma is treated by lowering intraocular pressure. Treatment options include prescription eye drops, oral medicines, laser treatment, surgery or a combination of approaches. What is cataract? According to Johns Hopkins Medicine, a cataract is a clouding of the lens of the eye.

As a cataract develops, your eyesight may become cloudy, blurry or unclear. You may experience Halos around lights, multiple vision and poor night vision. Colours may seem faded. How can cataract be treated? According to Cleveland Clinic, cataract surgery is the only way to remove cataracts and restore your clear vision. During cataract surgery and off film ologist removes your clouded natural lens and replaces it with an intraocular lens. IOL an IOL is an artificial lens that permanently stays in your eye. What is a Fibroid? According to Mayo Clinic, the thyroid is a small butterfly shaped gland located at the base of the neck just below the Adams apple. The thyroid gland makes 2 main hormones, Thyroxine T4 and three IODO Thyronine T3. These hormones affect every cell in the body. They support the rate at which the body uses fats and carbohydrates. How can thyroid be treated? Here's a summary from Mayo Clinic treatments, one anti Thyroid medicine, 2 beta blockers, 3 radio ID in therapy and four thyroidectomy. Can a person get thyroid in the stomach? According to NCBI, more distant location in which ectopic thyroid tissue has been reported include the heart, thymus, oesophagus, duodenum, gallbladder and adrenals. However, there are rare reports regarding ectopic thyroid in the stomach. What is a hernia? According to NHSA, hernia occurs when an internal part of the body pushes through a weakness in the muscle or surrounding tissue wall. A hernia usually develops between your chest and hips. In many cases, it causes no or very few symptoms, although you may notice a swelling or lump in your tummy, abdomen, or groyne. How can I hernia be treated?

According to NHS, inguinal hernias can be repaired using surgery to push the bulge back into place and strengthen the weakness in the abdominal wall. The operation is usually recommended if you have a hernia that causes pain, severe or persistent symptoms, or if any serious complications develop. What is the flu? According to Mayo Clinic, flu, also called influenza, is an infection of the nose, throat and lungs, which are part of the respiratory system. The flu is caused by a virus.

Influenza is commonly called the flu, but it's different from the stomach flu viruses that cause diarrhoea and vomiting. can Flus be treated? According to Mayo Clinic. But if you have a severe infection or retire risk of complications, your healthcare professional may prescribe an antiviral medicine to treat the flu. These medicines can include OSL, Tammy Beer, Tamiflu, Baloxavir, Exo Freezer and Zanamivir Relenza Arsenal. Tammy Veeran, barks Avira taken by mouth. What is whooping cough? According to Mayo Clinic, whooping cough, pertussis is a highly contagious respiratory tract infection. In many people. It's marked by a severe hacking cough followed by a high-pitched intake of breath that sounds like whoop. Before the vaccine was developed, whooping cough was considered a childhood disease. How can whooping cough be Cured? According to Centres for Disease Control and Prevention, doctors generally treat whooping cough with antibiotics. There are several antibiotics available to treat whooping cough. It's very important to treat whooping cough early, before coughing fits begin. Starting treatment after three weeks of illness is unlikely to help, even though most people will still have symptoms. What is Brain Fog? According to NHS Grampian, brain fog is not a medical term, but is used to describe a range of symptoms such as poor concentration, feeling confused, thinking more slowly than usual and fuzzy thoughts. What is Trumbo sis? According to CB thrombosis, the formation of a blood clot partial or complete blockage within blood vessels, whether venous or arterial, limiting the natural flow of blood and resulting in clinical sequala. How can Trumbo sis be cured? According to Penn Medicine, the standard of care for the treatment of acute DVT is blood thinning medication anticoagulation such as heparin and warfarin. Co made and blood thinning medications work by allowing blood to flow around a trapped clot while at the same time preventing clots from travelling to the lungs. What is a macular eye condition? According to Mayo Clinic vision with macular degeneration, dry macular degeneration is a common eye disorder among people over 50. It causes blurred or reduced central vision due to the breaking down of the inner layers of the macula, loo. The macula is the part of the retina that gives the I clear vision in the Direct Line of sight. What is a cure for the macular eye condition? According to NHS treatment depends on the type of AMD. You have Read about living with AMD wet AMD you make regular eye injection, Keighley a lighter dynamic therapy vision getting worse. Can a person have surgery for macular condition? According to Johns Hopkins Medicine, what is laser for you?

Literally Macule generation to coagulate attack surgery for the eyes. It is done to treat age-related macular degeneration. AMD is a condition that can lead to loss of vision. What is Diphtheria?

According to Centres for Disease Control and Prevention, Diphtheria? is a serious infection caused by strains of bacteria called Corrine bacterium diphtheria I that make toxin. It can lead to difficulty breathing, heart rhythm problems and even death. CDC recommends vaccines for infants, children, teens and adults to prevent diphtheria. Why do Children catch Chicken Pox? According to Kids Health, most kids with a sibling who's infected also will get it if they haven't already had the infection or the vaccine showing symptoms about two weeks after the first child does. Someone with chickenpox can spread the virus through droplets in the air by coughing or sneezing. Why do children catch measles? According to Centres for Disease Control and Prevention, child Diesel, just by being in a room where a person with measles has been even up to two hours after that person has left, an infected person can spread measles to others even before knowing, he/she has the disease- from four days before developing the measles rash through four days afterwards. How can a person catch shingles? According to New York State Department of Health, a person cannot get spell shingles from a person that has shingles. However, the virus that causes chickenpox and to be present with active shingles to a person who has never had chickenpox or the chickenpox vaccine.

What is malaria? According to Mayo Clinic, malaria is a disease caused by a parasite. The parasite is spread to humans through the bites of infected mosquitoes. People who have malaria usually feel very sick with a high fever and shaking chills. While the disease is uncommon in temperate climates, malaria is still common in tropical and subtropical countries. What is Gastroenteritis? According to Better Health Channel, so right shoulder infection and inflammation of the digestive system.

Symptoms can include our domino cramps, diarrhoea and vomiting. Some of the causes of gastroenteritis include viruses, bacteria, bacterial toxins, parasites, particular chemicals that can be treated, With pharmaceutical medication. What is a speech impediment? According to Maryville Online, each impediment is a condition that impacts an individual's ability to speak fluently, correctly, or with clear resonance or tone. Individuals with speech disorders have problem creating understandable sounds or forming words, leading to communication difficulties. Why does people get Eczema? According to Rene Testament, factors are out of children, dry weather and more specific things such as house dust, mite, animals, pollen and mould, food allergies such as allergies to cow's milk, eggs, peanuts, soya, wheat, certain materials worn next to the skin such as wool and synthetic fabrics. How can eczema be cured? According to NHS, there's no cure, but many children find their symptoms naturally improve as they get older. The main treatments for a topic Eczema emollients, moisturisers - used every day to stop the skin becoming dry, topical corticosteroids, creams and ointments used to swelling and redness during flare-ups. How do people get dermatitis? it can be caused by, some soaps and detergents, and some type of water allergy contact, dermatitis is less common and can be triggered by an allergic reaction to substances found in certain products, such as cosmetics or some metals, including nickel. What is Water allergy? According to the announcement, Can someone be allergic to water? It Seems like a strange question, but the answer is yes. There is a very rare condition known as aquagenic where skin comes in to contact with water, causes itchy red hives or swelling. In severe cases it can cause wheezing or shortness of breath.

How does coronavirus infect the body? According to MD, there is which is the proteins to receptors on healthy cells, especially those in your lungs. Specifically the viral proteins pass cells through a C2 receptors. Once inside the corona virus hijacks healthy cells and lodge in the body, to breakdown the cells. How does the coronavirus mutate in the body? According to OSF Healthcare, the hosts own cells, read the genetic code and replicate it, making more of the virus. That new virus then leaves the cell in search of another host to infect. Sometimes when that genetic code is being translated into proteins, a piece of the code gets changed. How can coronavirus be cured in the body? People can recover with Rapid body fluids, and the symptoms medicine you can get without a prescription can help, such as fever, reduces pain relievers such as ibuprofen or acetaminophen. How can we prevent ourselves from getting a coronavirus disease? Physically hygiene measures such as wearing face coverings to Can a coronavirus vaccine help to stop mutations in the body. What can be done to cure the coronavirus vaccine problem? How can the coronavirus vaccine prevent? Mutations . Correction Lower the rates by lowering the effective transmission rate. The effective transmission rates of single mutants given different coverage fractions of a completely effective vaccine for 100 percent, 70%, or 50% effective vaccine against transmission. entry requirements such as testing or proof of vaccination, state and three self isolation quarantine requirements. This book is a lifestyle prediction and a information guideline to help people understand, The body sickness, And get help, from medical professionals. A I technology, will predict the future of diagnostic test, On the human body, And for A I technology, Micro processer computers to send your medical situation reports to your doctors, And get medical professionals. To check your medical diagnostic report check-up. By A I Artificial Intelligence, Technology, could save a person's life. My research has been done. Through Google. Research. System. I've used my own knowledge, To write this book, With no help. From another person. On this day off: 27/3/2024 by patrick jackson.